OSTRICH FARMS

FUNKY FARMS

Lynn M. Stone

The Rourke Corporation, Inc.
Vero Beach, Florida 32964

PHOTO CREDITS:
All photos © Lynn M. Stone except page 4, 15 © Craig Lovell

EDITORIAL SERVICES:
Susan Albury

CREATIVE SERVICES:
East Coast Studios, Merritt Island, Florida

Library of Congress Cataloging-in-Publication Data

Stone, Lynn M.
  Ostrich farms / by Lynn M. Stone
    p. cm. — (Funky farms)
  Summary: Describes the physical characteristics and habits of ostriches and how these large birds are now being raised on farms across the United States.
  ISBN 0-86593-539-4
  1. Ostrich farms Juvenile literature. 2. Ostriches Juvenile literature. [1. Ostriches.
2. Ostrich farms.] I. Title. II. Series: Stone, Lynn M. Funky farms.
SF511.S76 1999
636.6'94—dc21   99-25302
                                                                              CIP

Printed in the USA

CONTENTS

OSTRICHES

The ostrich is a giant among birds. It stands nearly eight feet (2.4 meters) tall, almost a foot (30 centimeters) taller than the tallest NBA player! It can weigh nearly 350 pounds (160 kilograms).

The ostrich is not only the world's largest bird, it's one of the most unusual. It's the only bird with two toes on each foot.

The ostrich has plenty of feathers and two wings, like all birds. But the ostrich doesn't fly! Its small, floppy wings are useless for flight.

A male ostrich in black bounds across the grassy plains of Ngorongoro Crater in East Africa.

In the wild, the ostrich lives on the plains of Africa. There it runs for its life, rather than flies. With its long, powerful legs, an ostrich can race up to 40 miles per hour (65 kilometers per hour).

If an ostrich has to fight, it has a powerful kick.

Wild ostriches live mostly on plants. They also eat little **reptiles** (REP tilz), like lizards.

Two-toed feet are just one of the ostrich's unusual features. It is also the world's largest bird.

OSTRICH FARMS

Many ostrich farmers raise their birds as a hobby or as a part-time business. Some ostrich farmers, however, raise large numbers of ostriches as a full-time job.

Each year ostrich farmers in the United States **butcher** (BUHT chur), or kill, about 100,000 ostriches. Most of them are butchered between the ages of 10 and 14 months. Their meat, skins, and feathers are then sold by the farmers.

Ostriches produce meat, skins, and feathers for farmers. Visitors to ostrich farms can view the world's largest birds close-up.

HOW OSTRICH FARMS BEGAN

In the late 1800s, the long, lacy feathers of certain birds were in great demand. These feathers, called **plumes** (PLOOMZ), were worn on women's hats. To meet the demand for ostrich plumes, a few South Africans began ostrich farms.

During World War I (1914-1918), women's fashions changed. Feathers were out. But ostrich farms continued to sell meat, skins, and eggs. The feathers were sold as **feather dusters** (FEH thur DUHS turz) and decorations for homes.

Ostrich feathers are lacy, so they make ideal feather dusters. These dusters are for sale at an ostrich farm in California.

Farmers in the United States raise emus, too. Emus, from Australia, are also big, flightless birds. They are not, however, closely related to ostriches.

Ostriches do not bury their heads in the sand, as legend has it. But they do sleep with their necks flat on the ground, as this ostrich chick is doing.

WHERE OSTRICH FARMS ARE

Ostrich farms didn't become popular in the United States until the early 1980s. American farmers were more interested in selling the ostriches' meat and skin than its feathers.

Ostrich farms are scattered throughout the United States. They are also in northern and southern Africa, Australia, southern Europe, and Asia. South Africa raises the most ostriches.

Ostrich farmers keep older ostriches mainly in open areas, behind fences. The birds peck at the grass and other plants in their pasture. Most of their food, however, is given to them by the farmer.

Most ostrich farms are still in South Africa. Wide grasslands and a warm climate are ideal for the birds.

RAISING OSTRICHES

Ostrich food is dry and specially prepared by an animal food company. Ostriches grow quickly. During each of the first six months of their lives, ostriches can grow a foot (30 centimeters) taller.

From spring through summer, an ostrich hen will lay up to one egg every other day. The farmer usually removes each egg and places it in an **incubator** (INK u bay tur). If the farmer didn't remove the eggs, the hen would just lay 10 or 12 eggs and **incubate** (INK u bate) them herself.

This ostrich egg has pipped, which means the chick inside has begun to chip its way out.

Eggs must be kept warm. The incubator is a heated tray. It keeps air around the eggs at the perfect level of heat and dryness.

Ostrich eggs are the biggest of bird eggs. An ostrich egg may weigh three pounds (1.4 kilograms).

A baby ostrich hatches after six weeks in the incubator. The chick stands 10 inches (26 centimeters) tall at birth.

For three months, the young ostrich needs shelter and plenty of care. After three months, the young bird is less likely to become ill.

Those big ostrich eyes are nearly two inches (5 centimeters) across. The ostrich can spot danger a long distance away.

WHY OSTRICHES?

Ostrich meat is red, like beef, the meat of cattle. It has a beeflike taste, too, but ostrich meat has less fat than beef.

Most of the ostrich meat produced in the United States is sold in Europe.

Ostrich skins are valuable as leather. Ostrich leather is made into such products as boots, purses, belts, and jackets.

Ostrich feathers are divided into over 200 groups. Making them ready for sale is a hard job. Most of the best ostrich feathers come from South Africa.

Ostrich products include hand-painted ostrich eggs set on wood bases.

OSTRICH FARMS IN THE FUTURE

The market for ostrich steaks and burgers in the United States is growing. Many people are concerned about the amount of fat in their diets. That means that low-fat meats, such as ostrich, will probably become more popular. It's not likely, however, that Americans will begin serving ostrich on Thanksgiving.

Meanwhile, outside North America, ostrich farming is growing in Europe and Asia.

GLOSSARY

butcher (BUHT chur) — to kill an animal for its meat and for other human uses, such as leather

feather duster (FEH thur DUHS tur) — a group of lacy feathers, such as those of an ostrich, used to collect dust from furniture and other items

incubate (INK u bate) — to keep eggs in a condition favorable for hatching

incubator (INK u bay tur) — an instrument or place that keeps eggs or baby animals continually warm

plume (PLOOM) — a fancy, lacy bird feather, such as some of the feathers grown by ostriches and egrets

reptiles (REP tilz) — the group of animals including snakes, lizards, turtles, and the tuatara (too uh TAHR uh); cold-blooded, air-breathing animals with scaly skins or shells

INDEX

FURTHER READING

Find out more about ostriches with this helpful book and information site:
Stone, Lynn, *Ostriches*. Rourke, 1989

American Ostrich Association on line at www.ostriches.org